Twice Magic 5 by 5 squares

By Kermit Rose

ISBN-13:
978-1500970765

ISBN-10:
150097076X

Self Published on Create Space

5 by 5 magic squares

573	486	518	556	257
570	576	545	273	426
245	213	478	590	864
498	878	564	185	265
504	237	285	786	578

The 5 rows add to the magic number

573+486+518+556+257=2390

570+576+545+273+426=2390

245+213+478+590+864=2390

498+878+564+185+265=2390

504+237+285+786+578=2390

The 5 downward sloping diagonals add to the magic number.

573+576+478+185+578=2390

486+545+590+265+504=2390

518+273+864+498+237=2390

556+426+245+878+285=2390

257+570+213+564+786=2390

The 5 columns add to the magic number

573+570+245+498+504=2390

486+576+213+878+237=2390

518+545+478+564+285=2390

556+273+590+185+786=2390

257+426+864+265+578=2390

The 5 rows add to the magic number

573+486+518+556+257=2390

570+576+545+273+426=2390

245+213+478+590+864=2390

498+878+564+185+265=2390

504+237+285+786+578=2390

The 5 upward sloping diagonals add to the magic number.

573+237+564+590+426=2390

486+285+185+864+570=2390

518+786+265+245+576=2390

556+578+498+213+545=2390

504+878+478+273+257=2390

The above example magic square is derived by substituting
A=95 B=8   C=40 D=78 F=92 G=98 H=67
into the following formulas.

Row 1 column 1 = 2A+B+C+D+F+G+H
Row 1 column 2 = A+2B+C+D+F+G+H
Row 1 column 3 = A+B+2C+D+F+G+H
Row 1 column 4 = A+B+C+2D+F+G+H
Row 1 column 5 = F+G+H

Row 2 column 1 = A+B+C+D+2F+G+H
Row 2 column 2 = A+B+C+D+F+2G+H
Row 2 column 3 = A+B+C+D+F+G+2H
Row 2 column 4 = A+B+D+F
Row 2 column 5 = A+B+2C+D+G+H

Row 3 column 1= B+D+F+H
Row 3 column 2 = A+C+D
Row 3 column 3 = A+B+C+D+F+G+H
Row 3 column 4 = A+B+C+2F+2G+H
Row 3 column 5 = 2A+2B+2C+2D+F+2G+2H

Row 4 column 1 = A+B+C+F+2G+H
Row 4 column 2 = 2A+2B+2C+D+2F+2G+2H
Row 4 column 3 = A+2B+C+2D+F+G+H
Row 4 column 4 = C+D+H
Row 4 column 5 = A+D+F

Row 5 column 1 = A+B+2C+2D+G+H
Row 5 column 2 = D+F+H
Row 5 column 3 = A+F+G
Row 5 column 4 = 2A+2B+2C+D+F+2G+2H
Row 5 column 5 = A+2B+C+D+2F+G+H

The following 4 matrices show the twenty different ways to sum partitions of the magic square.

This first matrix is given to show that each column of the formula matrix sum to the same value.

The 1's show the first column; the 2's show the second column, etc.

1	2	3	4	5
1	2	3	4	5
1	2	3	4	5
1	2	3	4	5
1	2	3	4	5

This second matrix is given to show that each row of the formula matrix sum to the same value.

The 1's show the first row; the 2's show the second row, etc.

1	1	1	1	1
2	2	2	2	2
3	3	3	3	3
4	4	4	4	4
5	5	5	5	5

This third matrix is given to show that each downward sloping diagonal of the formula matrix sum to the same value.

The 1's show the first downward sloping diagonal; the 2's show the second downward diagonal, etc.

1	2	3	4	5
5	1	2	3	4
4	5	1	2	3
3	4	5	1	2
2	3	4	5	1

The fourth matrix is given to show that each upward sloping diagonal of the formula matrix sum to the same value.

The 5's show the principle upward sloping diagonal, the 4's show the next upward sloping diagonal, etc.

1	2	3	4	5
2	3	4	5	1
3	4	5	1	2
4	5	1	2	3
5	1	2	3	4

The following puzzles are based on the foregoing formulas.

For each puzzle, find the missing numbers that belong to the given magic square.

There are two ways to find the missing numbers.

[1] Find a row, column, or diagonal that contains exactly 4 numbers, and calculate the missing number.

[2] Use the foregoing formulas and the given numbers and solve for A, B, C, D, F, G, and H.

Then use the formulas for the cells with missing numbers to complete the square.

In some cases you may be able to mix these two methods.

You can even solve the puzzles by one method, and then come back later and solve by the other method.

Also, notice that the magic sum is always 5 times the number in the center, row 3, column 3.

Puzzle Number   1

12 Clue Numbers

Magic sum = 2390

573	xxx	xxx	xxx	xxx
570	xxx	xxx	273	426
xxx	xxx	478	590	864
498	878	xxx	xxx	xxx
504	237	xxx	xxx	578

Puzzle Number   2

14 Clue Numbers

Magic sum = 2335

548	473	xxx	564	206
xxx	xxx	xxx	274	454
xxx	255	xxx	533	xxx
443	837	xxx	xxx	xxx
551	230	xxx	747	563

Puzzle Number   3

12 Clue Numbers

Magic sum = 1540

337	315	xxx	370	xxx
328	xxx	xxx	xxx	xxx
113	xxx	xxx	xxx	596
xxx	554	xxx	156	111
420	106	xxx	534	xxx

Puzzle Number   4

14 Clue Numbers

Magic sum = 1410

347	302	xxx	xxx	122
xxx	xxx	xxx	88	xxx
52	140	xxx	374	562
372	563	xxx	104	68
xxx	32	xxx	xxx	304

Puzzle Number   5

13 Clue Numbers

Magic sum = 2150

499	xxx	xxx	xxx	xxx
469	xxx	xxx	196	466
194	207	xxx	498	821
xxx	797	xxx	205	171
529	xxx	200	xxx	xxx

Puzzle Number   6

14 Clue Numbers

Magic sum = 615

123	140	xxx	135	54
xxx	xxx	xxx	55	xxx
xxx	52	xxx	156	xxx
130	234	152	61	38
149	xxx	xxx	xxx	166

Puzzle Number   7

13 Clue Numbers

Magic sum = 1195

253	xxx	282	xxx	161
xxx	273	xxx	xxx	228
xxx	xxx	xxx	327	424
xxx	478	xxx	116	xxx
228	xxx	102	424	314

Puzzle Number   8

12 Clue Numbers

Magic sum = 1625

366	xxx	xxx	xxx	xxx
xxx	382	xxx	xxx	330
170	xxx	xxx	418	xxx
366	634	xxx	158	xxx
346	xxx	150	582	394

Puzzle Number   9

13 Clue Numbers

Magic sum = 1860

xxx	xxx	401	463	98
xxx	402	xxx	xxx	357
xxx	xxx	372	355	xxx
311	xxx	xxx	144	xxx
448	xxx	158	609	486

Puzzle Number   10

13 Clue Numbers

Magic sum = 2385

563	548	xxx	xxx	xxx
xxx	xxx	575	xxx	453
xxx	187	477	569	858
473	xxx	xxx	xxx	xxx
482	223	207	829	xxx

Puzzle Number   11

13 Clue Numbers

Magic sum = 2120

523	442	xxx	xxx	xxx
485	518	480	254	xxx
xxx	xxx	424	xxx	787
442	xxx	xxx	xxx	xxx
459	193	254	xxx	503

Puzzle Number   12

12 Clue Numbers

Magic sum = 1885

412	xxx	460	xxx	128
xxx	468	xxx	xxx	457
168	161	xxx	428	751
xxx	xxx	xxx	xxx	xxx
500	80	xxx	xxx	468

Puzzle Number 13

14 Clue Numbers

Magic sum = 1840

440	417	368	xxx	xxx
xxx	404	455	xxx	307
260	135	xxx	xxx	xxx
341	673	xxx	xxx	xxx
370	211	xxx	612	478

Puzzle Number 14

13 Clue Numbers

Magic sum = 2040

418	xxx	475	xxx	149
xxx	xxx	501	xxx	xxx
315	xxx	408	xxx	776
339	xxx	590	xxx	135
520	xxx	xxx	691	545

Puzzle Number   15

13 Clue Numbers

Magic sum = 1690

406	xxx	413	xxx	95
xxx	346	xxx	229	xxx
xxx	215	338	xxx	xxx
274	xxx	438	xxx	xxx
424	159	137	xxx	427

Puzzle Number   16

13 Clue Numbers

Magic sum = 2095

451	516	xxx	xxx	227
xxx	xxx	xxx	xxx	xxx
275	xxx	419	539	767
468	xxx	xxx	xxx	136
411	178	185	xxx	587

Puzzle Number   17

15 Clue Numbers

Magic sum = 1955

xxx	xxx	xxx	436	xxx
xxx	436	451	243	xxx
xxx	140	391	473	700
xxx	737	500	148	179
397	xxx	xxx	655	537

Puzzle Number   18

13 Clue Numbers

Magic sum = 1650

380	xxx	xxx	347	155
xxx	xxx	xxx	178	327
221	xxx	330	xxx	612
327	643	410	155	xxx
xxx	xxx	xxx	595	xxx

Puzzle Number 19

12 Clue Numbers

Magic sum = 1930

xxx	482	xxx	xxx	xxx
xxx	391	431	288	412
xxx	218	386	317	750
xxx	676	578	189	xxx
xxx	xxx	xxx	xxx	xxx

Puzzle Number 20

12 Clue Numbers

Magic sum = 1655

xxx	xxx	xxx	xxx	xxx
xxx	377	xxx	195	329
241	xxx	331	333	626
297	xxx	480	170	xxx
xxx	xxx	xxx	546	436

Puzzle Number 21

13 Clue Numbers

Magic sum = 1170

xxx	xxx	xxx	xxx	xxx
xxx	250	301	58	306
xxx	129	234	xxx	447
250	xxx	xxx	160	57
xxx	88	73	xxx	256

Puzzle Number 22

12 Clue Numbers

Magic sum = 1525

356	xxx	xxx	353	xxx
323	xxx	xxx	164	xxx
xxx	165	305	337	xxx
xxx	562	xxx	127	xxx
xxx	79	131	xxx	370

Puzzle Number   23

12 Clue Numbers

Magic sum = 1725

351	xxx	xxx	xxx	204
xxx	xxx	373	xxx	xxx
xxx	122	345	488	607
405	xxx	397	xxx	xxx
xxx	144	xxx	574	447

Puzzle Number   24

13 Clue Numbers

Magic sum = 1635

381	xxx	xxx	363	xxx
348	xxx	356	xxx	xxx
140	xxx	327	406	633
xxx	xxx	xxx	104	111
381	xxx	xxx	597	402

Puzzle Number   25

13 Clue Numbers

Magic sum = 1755

418	398	368	358	xxx
xxx	411	xxx	xxx	274
xxx	91	xxx	498	xxx
404	695	xxx	xxx	xxx
281	xxx	xxx	601	492

Puzzle Number   26

14 Clue Numbers

Magic sum = 1340

286	xxx	xxx	296	xxx
331	324	314	xxx	241
xxx	82	268	xxx	473
xxx	508	xxx	xxx	xxx
269	137	xxx	445	352

Puzzle Number   27

13 Clue Numbers

Magic sum = 1965

433	421	463	xxx	xxx
444	491	xxx	xxx	412
185	xxx	393	471	xxx
420	715	492	xxx	xxx
xxx	xxx	xxx	664	xxx

Puzzle Number   28

12 Clue Numbers

Magic sum = 1325

286	xxx	304	xxx	xxx
275	299	xxx	xxx	xxx
xxx	105	265	xxx	xxx
254	485	367	xxx	76
339	xxx	65	xxx	xxx

Puzzle Number   29

13 Clue Numbers

Magic sum = 2010

477	463	475	xxx	161
xxx	xxx	470	xxx	xxx
xxx	180	402	463	723
xxx	772	xxx	xxx	188
426	xxx	xxx	xxx	544

Puzzle Number   30

13 Clue Numbers

Magic sum = 1405

xxx	xxx	285	333	127
368	xxx	317	237	198
xxx	xxx	281	xxx	475
xxx	xxx	xxx	xxx	xxx
250	xxx	176	423	381

Puzzle Number   31

13 Clue Numbers

Magic sum = 1855

xxx	407	xxx	xxx	xxx
416	xxx	418	xxx	355
172	xxx	371	xxx	697
411	xxx	451	xxx	175
399	xxx	215	xxx	452

Puzzle Number   32

13 Clue Numbers

Magic sum = 1010

208	245	244	xxx	xxx
xxx	xxx	228	97	212
xxx	xxx	202	255	372
xxx	xxx	261	84	54
xxx	xxx	xxx	xxx	277

Puzzle Number 33

13 Clue Numbers

Magic sum = 2400

xxx	xxx	546	554	237
542	559	xxx	xxx	xxx
xxx	xxx	480	xxx	898
485	xxx	xxx	236	148
558	232	xxx	xxx	633

Puzzle Number 34

13 Clue Numbers

Magic sum = 2085

484	xxx	480	506	xxx
xxx	504	472	xxx	xxx
200	xxx	417	xxx	823
415	xxx	xxx	xxx	xxx
xxx	155	165	734	473

Puzzle Number 35

13 Clue Numbers

Magic sum = 1800

415	419	381	xxx	206
xxx	xxx	xxx	xxx	303
xxx	95	360	xxx	642
xxx	xxx	438	137	xxx
322	xxx	164	xxx	497

Puzzle Number 36

13 Clue Numbers

Magic sum = 1265

352	xxx	257	xxx	99
261	285	xxx	xxx	249
xxx	110	253	xxx	xxx
xxx	499	304	xxx	xxx
256	xxx	139	xxx	305

Puzzle Number 37

12 Clue Numbers

Magic sum = 1865

383	xxx	xxx	437	xxx
445	xxx	452	xxx	xxx
288	97	xxx	xxx	xxx
xxx	682	xxx	166	146
388	215	134	xxx	xxx

Puzzle Number 38

13 Clue Numbers

Magic sum = 1580

387	364	xxx	339	151
xxx	xxx	335	188	xxx
136	xxx	xxx	425	586
xxx	609	387	xxx	xxx
316	88	xxx	xxx	xxx

Puzzle Number   39

14 Clue Numbers

Magic sum = 1370

287	xxx	332	xxx	147
309	310	350	xxx	xxx
xxx	103	xxx	313	xxx
xxx	xxx	330	166	80
329	xxx	84	481	xxx

Puzzle Number   40

13 Clue Numbers

Magic sum = 2220

528	467	xxx	533	xxx
465	xxx	xxx	217	473
212	223	xxx	474	xxx
xxx	xxx	556	218	xxx
xxx	189	203	xxx	xxx

Puzzle Number   41

13 Clue Numbers

Magic sum = 2290

xxx	xxx	517	490	225
xxx	xxx	xxx	267	424
222	190	458	597	xxx
xxx	xxx	533	145	224
xxx	xxx	270	xxx	xxx

Puzzle Number   42

12 Clue Numbers

Magic sum = 1480

322	371	xxx	299	xxx
xxx	356	xxx	xxx	212
206	33	296	xxx	xxx
xxx	xxx	xxx	xxx	117
215	xxx	174	xxx	459

Puzzle Number   43

12 Clue Numbers

Magic sum = 1695

xxx	xxx	xxx	345	xxx
371	xxx	xxx	xxx	339
xxx	124	xxx	xxx	646
xxx	672	400	101	124
xxx	101	183	xxx	426

Puzzle Number   44

12 Clue Numbers

Magic sum = 1410

xxx	xxx	367	xxx	128
308	xxx	xxx	95	xxx
xxx	xxx	xxx	xxx	538
xxx	518	345	174	xxx
xxx	115	91	492	325

Puzzle Number   45

14 Clue Numbers

Magic sum = 2220

524	xxx	xxx	482	141
xxx	xxx	513	257	490
xxx	205	444	xxx	xxx
437	850	xxx	194	xxx
xxx	xxx	152	809	583

Puzzle Number   46

14 Clue Numbers

Magic sum = 1440

306	xxx	312	xxx	145
xxx	294	xxx	xxx	222
xxx	113	288	313	xxx
223	505	xxx	144	xxx
293	xxx	114	415	xxx

Puzzle Number   47

13 Clue Numbers

Magic sum = 1735

366	xxx	439	xxx	xxx
368	365	xxx	166	xxx
xxx	xxx	347	301	673
xxx	609	xxx	xxx	125
503	177	xxx	588	xxx

Puzzle Number   48

14 Clue Numbers

Magic sum = 1575

xxx	367	xxx	322	159
348	390	366	xxx	xxx
xxx	xxx	315	416	597
383	xxx	374	63	xxx
294	91	xxx	xxx	xxx

Puzzle Number   49

16 Clue Numbers

Magic sum = 1080

218	233	227	xxx	xxx
xxx	289	257	xxx	221
130	xxx	xxx	229	426
xxx	xxx	299	118	74
287	113	xxx	360	239

Puzzle Number   50

14 Clue Numbers

Magic sum = 1355

273	xxx	xxx	xxx	xxx
xxx	310	xxx	xxx	xxx
210	62	271	xxx	455
270	502	387	xxx	129
244	134	128	415	xxx

Puzzle Number   51

13 Clue Numbers

Magic sum = 1195

279	321	xxx	xxx	xxx
298	270	246	xxx	xxx
153	xxx	239	xxx	xxx
265	473	326	xxx	xxx
xxx	71	130	414	xxx

Puzzle Number   52

14 Clue Numbers

Magic sum = 1430

xxx	287	xxx	375	152
287	375	xxx	xxx	302
153	xxx	286	xxx	571
xxx	xxx	376	168	xxx
391	152	xxx	482	xxx

Puzzle Number 53

13 Clue Numbers

Magic sum = 1500

xxx	345	350	xxx	xxx
323	390	314	xxx	xxx
139	128	300	356	xxx
333	543	402	xxx	xxx
384	xxx	xxx	xxx	xxx

Puzzle Number 54

15 Clue Numbers

Magic sum = 1825

383	414	444	xxx	120
xxx	xxx	415	177	xxx
209	196	xxx	336	719
xxx	631	xxx	228	xxx
532	xxx	88	xxx	425

Puzzle Number   55

13 Clue Numbers

Magic sum = 1720

xxx	xxx	xxx	353	184
xxx	396	xxx	xxx	358
xxx	142	344	xxx	644
xxx	xxx	371	155	128
xxx	xxx	171	635	406

Puzzle Number   56

14 Clue Numbers

Magic sum = 1490

368	xxx	xxx	xxx	xxx
xxx	xxx	345	206	244
183	xxx	298	349	xxx
268	548	353	122	199
292	xxx	xxx	xxx	386

Puzzle Number   57

13 Clue Numbers

Magic sum = 1295

xxx	xxx	273	xxx	108
xxx	xxx	262	216	xxx
xxx	67	xxx	329	439
xxx	483	378	52	132
229	xxx	xxx	xxx	422

Puzzle Number   58

12 Clue Numbers

Magic sum = 1500

xxx	xxx	xxx	xxx	xxx
xxx	xxx	392	161	xxx
243	61	300	355	xxx
xxx	550	xxx	143	119
xxx	xxx	115	491	401

Puzzle Number   59

12 Clue Numbers

Magic sum = 2070

xxx	xxx	xxx	497	xxx
435	xxx	xxx	263	xxx
xxx	172	414	xxx	xxx
391	745	xxx	xxx	186
483	xxx	163	724	512

Puzzle Number   60

13 Clue Numbers

Magic sum = 1645

387	392	xxx	366	xxx
xxx	xxx	xxx	208	359
xxx	175	329	344	xxx
xxx	621	429	xxx	xxx
xxx	126	110	xxx	442

Puzzle Number   61

12 Clue Numbers

Magic sum = 1485

363	xxx	xxx	xxx	118
351	323	xxx	231	245
xxx	88	297	357	540
xxx	xxx	xxx	xxx	xxx
265	112	xxx	xxx	xxx

Puzzle Number   62

13 Clue Numbers

Magic sum = 1850

461	xxx	xxx	446	125
xxx	395	xxx	289	xxx
245	176	xxx	372	xxx
xxx	xxx	515	xxx	220
402	xxx	xxx	611	492

Puzzle Number 63

13 Clue Numbers

Magic sum = 1870

xxx	xxx	395	402	xxx
xxx	382	xxx	283	xxx
xxx	126	xxx	438	664
xxx	720	xxx	111	189
339	xxx	169	xxx	552

Puzzle Number 64

13 Clue Numbers

Magic sum = 1310

xxx	xxx	279	xxx	155
xxx	342	xxx	119	250
121	89	xxx	xxx	495
314	xxx	308	91	xxx
278	xxx	xxx	467	xxx

Puzzle Number   65

12 Clue Numbers

Magic sum = 1060

xxx	xxx	xxx	271	59
xxx	xxx	xxx	126	239
xxx	xxx	212	158	423
157	365	326	xxx	xxx
xxx	114	16	xxx	xxx

Puzzle Number   66

14 Clue Numbers

Magic sum = 1835

xxx	400	462	xxx	xxx
465	xxx	xxx	168	364
xxx	132	367	487	xxx
389	713	xxx	xxx	135
385	xxx	xxx	615	498

Puzzle Number   67

13 Clue Numbers

Magic sum = 2035

xxx	xxx	421	450	218
458	477	504	xxx	xxx
xxx	103	407	xxx	xxx
434	771	xxx	xxx	140
413	191	xxx	xxx	xxx

Puzzle Number   68

12 Clue Numbers

Magic sum = 1655

xxx	xxx	392	341	146
xxx	392	340	xxx	xxx
185	95	xxx	xxx	xxx
382	652	xxx	xxx	110
xxx	95	161	xxx	xxx

Puzzle Number 69

14 Clue Numbers

Magic sum = 2235

502	508	xxx	xxx	xxx
xxx	xxx	509	204	542
211	238	xxx	xxx	893
445	807	595	xxx	xxx
xxx	150	141	806	xxx

Puzzle Number 70

13 Clue Numbers

Magic sum = 1945

478	xxx	443	xxx	183
xxx	xxx	410	xxx	xxx
157	xxx	389	xxx	xxx
xxx	772	xxx	81	xxx
376	100	251	699	519

Puzzle Number 71

13 Clue Numbers

Magic sum = 1995

459	xxx	xxx	430	198
492	xxx	xxx	xxx	xxx
301	xxx	399	481	xxx
xxx	xxx	xxx	134	184
355	209	173	xxx	584

Puzzle Number 72

13 Clue Numbers

Magic sum = 1850

393	446	437	xxx	186
440	xxx	xxx	187	367
xxx	xxx	370	xxx	670
402	xxx	xxx	xxx	xxx
385	xxx	143	652	xxx

Puzzle Number   73

12 Clue Numbers

Magic sum = 1620

xxx	380	xxx	xxx	117
xxx	340	xxx	261	270
xxx	xxx	324	xxx	575
262	xxx	xxx	125	xxx
348	179	xxx	xxx	453

Puzzle Number   74

13 Clue Numbers

Magic sum = 2215

531	542	xxx	452	xxx
xxx	500	xxx	289	436
xxx	183	443	xxx	793
xxx	xxx	551	xxx	xxx
445	xxx	xxx	784	635

Puzzle Number   75

13 Clue Numbers

Magic sum = 1185

xxx	xxx	250	249	141
333	257	262	xxx	xxx
186	xxx	237	341	378
xxx	462	xxx	xxx	126
166	xxx	xxx	xxx	xxx

Puzzle Number   76

12 Clue Numbers

Magic sum =   2200

452	xxx	xxx	xxx	268
xxx	xxx	530	166	xxx
244	115	440	xxx	xxx
521	xxx	xxx	193	109
xxx	xxx	xxx	783	578

Puzzle Number   77

12 Clue Numbers

Magic sum = 1370

xxx	xxx	288	xxx	xxx
xxx	xxx	xxx	233	228
157	xxx	274	357	xxx
297	xxx	369	18	140
xxx	64	xxx	486	xxx

Puzzle Number   78

14 Clue Numbers

Magic sum = 1480

395	332	392	xxx	63
xxx	296	344	xxx	377
101	197	xxx	xxx	xxx
xxx	xxx	334	146	xxx
379	xxx	xxx	575	347

Puzzle Number   79

13 Clue Numbers

Magic sum = 1520

xxx	322	xxx	xxx	xxx
xxx	380	xxx	131	332
98	141	304	xxx	564
xxx	xxx	333	108	113
343	xxx	xxx	xxx	366

Puzzle Number   80

12 Clue Numbers

Magic sum = 1735

429	xxx	348	xxx	103
405	350	xxx	301	xxx
xxx	149	xxx	342	xxx
xxx	xxx	508	109	xxx
356	xxx	xxx	xxx	500

Puzzle Number   81

14 Clue Numbers

Magic sum = 1940

xxx	xxx	394	482	173
xxx	xxx	xxx	xxx	324
267	129	388	xxx	706
380	682	568	xxx	xxx
418	181	xxx	612	xxx

Puzzle Number   82

14 Clue Numbers

Magic sum = 1700

356	xxx	391	xxx	183
388	xxx	396	154	xxx
194	110	340	xxx	632
xxx	637	xxx	xxx	107
xxx	147	xxx	589	xxx

Puzzle Number 83

13 Clue Numbers

Magic sum = 2395

562	xxx	xxx	575	xxx
xxx	541	xxx	242	xxx
258	255	479	xxx	xxx
xxx	862	586	xxx	231
599	xxx	xxx	810	542

Puzzle Number 84

12 Clue Numbers

Magic sum = 1355

xxx	xxx	xxx	xxx	xxx
305	292	293	216	xxx
168	137	271	271	xxx
xxx	487	xxx	89	xxx
304	xxx	xxx	xxx	362

Puzzle Number   85

12 Clue Numbers

Magic sum = 2250

473	522	xxx	xxx	179
513	xxx	526	xxx	xxx
xxx	199	450	xxx	837
xxx	xxx	xxx	xxx	180
563	xxx	126	743	xxx

Puzzle Number   86

12 Clue Numbers

Magic sum = 1210

xxx	243	xxx	292	xxx
277	xxx	262	xxx	240
xxx	109	xxx	xxx	xxx
269	434	xxx	103	xxx
290	xxx	138	399	xxx

Puzzle Number   87

15 Clue Numbers

Magic sum = 980

xxx	227	xxx	220	47
xxx	xxx	214	125	220
74	xxx	196	201	391
200	xxx	xxx	xxx	xxx
244	43	98	xxx	228

Puzzle Number   88

13 Clue Numbers

Magic sum = 2045

xxx	502	xxx	490	xxx
xxx	xxx	xxx	258	405
xxx	xxx	409	471	766
xxx	737	583	xxx	165
486	145	175	xxx	xxx

Puzzle Number 89

12 Clue Numbers

Magic sum = 1655

xxx	xxx	xxx	401	xxx
341	404	xxx	161	386
xxx	xxx	331	344	xxx
334	xxx	xxx	167	113
456	xxx	xxx	xxx	389

Puzzle Number 90

12 Clue Numbers

Magic sum = 1980

477	453	471	xxx	xxx
419	419	485	209	xxx
xxx	204	xxx	394	xxx
371	xxx	xxx	xxx	152
xxx	xxx	xxx	xxx	476

Puzzle Number   91

13 Clue Numbers

Magic sum = 1480

xxx	297	376	xxx	xxx
343	xxx	308	119	xxx
xxx	xxx	296	400	xxx
353	564	xxx	xxx	118
xxx	87	175	517	xxx

Puzzle Number   92

14 Clue Numbers

Magic sum = 1085

253	261	xxx	258	70
xxx	xxx	xxx	128	xxx
94	xxx	217	xxx	427
237	xxx	xxx	xxx	84
xxx	50	104	386	268

Puzzle Number 93

12 Clue Numbers

Magic sum = 1775

xxx	xxx	xxx	377	160
xxx	xxx	398	xxx	398
xxx	116	355	xxx	669
409	688	456	xxx	xxx
420	xxx	127	xxx	xxx

Puzzle Number 94

13 Clue Numbers

Magic sum = 1565

371	xxx	388	391	100
xxx	xxx	xxx	xxx	xxx
111	xxx	313	xxx	609
304	548	xxx	xxx	xxx
449	109	xxx	531	332

Puzzle Number   95

12 Clue Numbers

Magic sum = 1895

xxx	469	xxx	463	193
419	xxx	440	xxx	xxx
xxx	xxx	379	xxx	718
387	674	xxx	xxx	xxx
423	185	xxx	xxx	509

Puzzle Number   96

13 Clue Numbers

Magic sum = 1315

281	xxx	286	271	xxx
xxx	350	xxx	146	xxx
xxx	49	263	408	xxx
342	518	xxx	xxx	92
228	xxx	xxx	xxx	383

Puzzle Number   97

14 Clue Numbers

Magic sum = 1565

313	390	351	xxx	xxx
xxx	xxx	326	192	290
205	92	xxx	xxx	565
xxx	572	xxx	105	xxx
xxx	xxx	131	511	451

Puzzle Number   98

13 Clue Numbers

Magic sum = 2155

xxx	470	530	xxx	xxx
517	xxx	490	xxx	xxx
184	xxx	xxx	586	776
500	862	xxx	xxx	165
xxx	145	234	xxx	556

Puzzle Number   99

13 Clue Numbers

Magic sum = 1635

401	xxx	396	xxx	xxx
399	xxx	345	xxx	xxx
xxx	228	327	xxx	xxx
249	569	414	xxx	231
xxx	175	xxx	497	401

Solution   1

573	486	518	556	257
570	576	545	273	426
245	213	478	590	864
498	878	564	185	265
504	237	285	786	578

Solution   2

548	473	544	564	206
557	540	510	274	454
236	255	467	533	844
443	837	570	217	268
551	230	244	747	563

Solution   3

337	315	378	370	140
328	404	332	118	358
113	161	308	362	596
342	554	377	156	111
420	106	145	534	335

Solution   4

347	302	356	283	122
284	373	311	88	354
52	140	282	374	562
372	563	303	104	68
355	32	158	561	304

Solution   5

499	455	505	493	198
469	522	497	196	466
194	207	430	498	821
459	797	518	205	171
529	169	200	758	494

Solution   6

123	140	163	135	54
149	142	132	55	137
64	52	123	156	220
130	234	152	61	38
149	47	45	208	166

Solution 7

253	260	282	239	161
293	273	312	89	228
148	57	239	327	424
273	478	260	116	68
228	127	102	424	314

Solution 8

366	342	382	341	194
377	382	410	126	330
170	114	325	418	598
366	634	358	158	109
346	153	150	582	394

Solution 9

456	442	401	463	98
416	402	396	289	357
229	204	372	355	700
311	653	533	144	219
448	159	158	609	486

Solution   10

563	548	549	506	219
573	502	575	282	453
294	187	477	569	858
473	925	577	199	211
482	223	207	829	644

Solution   11

523	442	444	500	211
485	518	480	254	383
211	195	424	503	787
442	772	518	152	236
459	193	254	711	503

Solution   12

412	465	460	420	128
380	468	411	169	457
168	161	377	428	751
425	711	508	160	81
500	80	129	708	468

Solution 13

440	417	368	431	184
429	404	455	245	307
260	135	368	402	675
341	673	480	150	196
370	211	169	612	478

Solution 14

418	505	475	493	149
448	424	501	232	435
315	162	408	379	776
339	731	590	245	135
520	218	66	691	545

Solution 15

406	366	413	410	95
399	346	364	229	352
187	215	338	335	615
274	604	438	173	201
424	159	137	543	427

Solution   16

451	516	449	452	227
490	501	493	233	378
275	95	419	539	767
468	805	549	137	136
411	178	185	734	587

Solution   17

443	455	434	436	187
473	436	451	243	352
251	140	391	473	700
391	737	500	148	179
397	187	179	655	537

Solution   18

380	393	375	347	155
378	344	423	178	327
221	112	330	375	612
327	643	410	155	115
344	158	112	595	441

Solution 19

460	482	434	482	72
408	391	431	288	412
259	218	386	317	750
295	676	578	189	192
508	163	101	654	504

Solution 20

341	400	365	411	138
367	377	387	195	329
241	124	331	333	626
297	582	480	170	126
409	172	92	546	436

Solution 21

270	235	327	234	104
255	250	301	58	306
89	129	234	271	447
250	468	235	160	57
306	88	73	447	256

Solution   22

356	352	371	353	93
323	367	318	164	353
126	165	305	337	592
319	562	400	127	117
401	79	131	544	370

Solution   23

351	364	428	378	204
428	438	373	141	345
163	122	345	488	607
405	657	397	144	122
378	144	182	574	447

Solution   24

381	381	366	363	144
348	421	356	165	345
140	129	327	406	633
385	618	417	104	111
381	86	169	597	402

Solution   25

418	398	368	358	213
445	411	410	215	274
207	91	351	498	608
404	695	405	83	168
281	160	221	601	492

Solution   26

286	289	304	296	165
331	324	314	130	241
158	82	268	359	473
296	508	317	110	109
269	137	137	445	352

Solution   27

433	421	463	464	184
444	491	428	190	412
185	181	393	471	735
420	715	492	176	162
483	157	189	664	472

Solution 28

286	322	304	310	103
275	299	324	133	294
171	105	265	264	520
254	485	367	143	76
339	114	65	475	332

Solution 29

477	463	475	434	161
483	414	470	249	394
242	180	402	463	723
382	772	495	173	188
426	181	168	691	544

Solution 30

366	294	285	333	127
368	285	317	237	198
188	141	281	320	475
233	510	346	92	224
250	175	176	423	381

Solution   31

457	407	400	415	176
416	455	418	211	355
172	159	371	456	697
411	698	451	120	175
399	136	215	653	452

Solution   32

208	245	244	218	95
234	239	228	97	212
117	64	202	255	372
223	388	261	84	54
228	74	75	356	277

Solution   33

492	571	546	554	237
542	559	576	239	484
323	152	480	547	898
485	886	645	236	148
558	232	153	824	633

Solution   34

484	462	480	506	153
428	504	472	212	469
200	219	417	426	823
415	745	551	207	167
558	155	165	734	473

Solution   35

415	419	381	379	206
438	391	457	211	303
253	95	360	450	642
372	701	438	137	152
322	194	164	623	497

Solution   36

352	297	257	260	99
261	285	312	158	249
118	110	253	286	498
278	499	304	70	114
256	74	139	491	305

Solution 37

383	446	396	437	203
445	425	452	219	324
288	97	373	433	674
361	682	510	166	146
388	215	134	610	518

Solution 38

387	364	339	339	151
362	402	335	188	293
136	117	316	425	586
379	609	387	65	140
316	88	203	563	410

Solution 39

287	298	332	306	147
309	310	350	104	297
167	103	274	313	513
278	516	330	166	80
329	143	84	481	333

Solution   40

528	467	494	533	198
465	542	523	217	473
212	223	444	474	867
453	799	556	218	194
562	189	203	778	488

Solution   41

557	501	517	490	225
551	536	512	267	424
222	190	458	597	823
504	884	533	145	224
456	179	270	791	594

Solution   42

322	371	300	299	188
384	356	336	192	212
206	33	296	441	504
353	589	374	47	117
215	131	174	501	459

Solution 43

425	394	371	345	160
371	404	402	179	339
156	124	339	430	646
398	672	400	101	124
345	101	183	640	426

Solution 44

288	299	367	328	128
308	341	325	95	341
132	137	282	321	538
295	518	345	174	78
387	115	91	492	325

Solution 45

524	542	531	482	141
485	475	513	257	490
246	205	444	478	847
437	850	580	194	159
528	148	152	809	583

Solution 46

306	318	312	359	145
378	294	337	209	222
240	113	288	313	486
223	505	389	144	179
293	210	114	415	408

Solution 47

366	388	439	432	110
368	365	418	166	418
218	196	347	301	673
280	609	473	248	125
503	177	58	588	409

Solution 48

407	367	320	322	159
348	390	366	184	287
143	104	315	416	597
383	623	374	63	132
294	91	200	590	400

Solution 49

218	233	227	282	120
222	289	257	91	221
130	79	216	229	426
223	366	299	118	74
287	113	81	360	239

Solution 50

273	347	291	311	133
358	310	278	205	204
210	62	271	357	455
270	502	387	67	129
244	134	128	415	434

Solution 51

279	321	254	244	97
298	270	246	186	195
153	60	239	324	419
265	473	326	27	104
200	71	130	414	380

Solution   52

313	287	303	375	152
287	375	348	118	302
153	133	286	287	571
286	483	376	168	117
391	152	117	482	288

Solution   53

321	345	350	357	127
323	390	314	146	327
139	128	300	356	577
333	543	402	121	101
384	94	134	520	368

Solution   54

383	414	444	464	120
376	424	415	177	433
209	196	365	336	719
325	631	513	228	128
532	160	88	620	425

Solution   55

419	362	402	353	184
388	396	432	146	358
159	142	344	431	644
387	679	371	155	128
367	141	171	635	406

Solution   56

368	305	325	346	146
379	316	345	206	244
183	145	298	349	515
268	548	353	122	199
292	176	169	467	386

Solution   57

277	343	273	294	108
338	285	262	216	194
201	67	259	329	439
250	483	378	52	132
229	117	123	404	422

Solution   58

310	342	301	350	197
359	346	392	161	242
243	61	300	355	541
296	550	392	143	119
292	201	115	491	401

Solution   59

496	491	421	497	165
435	474	498	263	400
265	172	414	412	807
391	745	574	174	186
483	188	163	724	512

Solution   60

387	392	409	366	91
379	331	368	208	359
189	175	329	344	608
294	621	429	156	145
396	126	110	571	442

Solution 61

363	388	299	317	118
351	323	335	231	245
203	88	297	357	540
303	574	408	60	140
265	112	146	520	442

Solution 62

461	439	379	446	125
423	395	417	289	326
245	176	370	372	687
319	664	515	132	220
402	176	169	611	492

Solution 63

451	468	395	402	154
458	382	436	283	311
268	126	374	438	664
354	720	496	111	189
339	174	169	636	552

Solution 64

306	280	279	290	155
291	342	308	119	250
121	89	262	343	495
314	496	308	91	101
278	103	153	467	309

Solution 65

223	267	240	271	59
213	216	266	126	239
169	98	212	158	423
157	365	326	141	71
298	114	16	364	268

Solution 66

383	400	462	388	202
465	410	428	168	364
213	132	367	487	636
389	713	421	177	135
385	180	157	615	498

Solution   67

453	493	421	450	218
458	477	504	226	370
277	103	407	485	763
434	771	536	154	140
413	191	167	720	544

Solution   68

355	421	392	341	146
407	392	340	200	316
185	95	331	458	586
382	652	431	80	110
326	95	161	576	497

Solution   69

502	508	543	534	148
448	532	509	204	542
211	238	447	446	893
445	807	595	245	143
629	150	141	806	509

Solution   70

478	446	443	395	183
462	478	410	225	370
157	149	389	545	705
472	772	452	81	168
376	100	251	699	519

Solution   71

459	491	417	430	198
492	419	484	276	324
301	109	399	481	705
388	767	522	134	184
355	209	173	674	584

Solution   72

393	446	437	388	186
440	420	436	187	367
230	108	370	472	670
402	722	464	151	111
385	154	143	652	516

Solution   73

378	380	343	402	117
397	340	352	261	270
235	151	324	335	575
262	570	458	125	205
348	179	143	497	453

Solution   74

531	542	529	452	161
536	500	454	289	436
212	183	443	584	793
491	877	551	106	190
445	113	238	784	635

Solution   75

255	290	250	249	141
333	257	262	179	154
186	43	237	341	378
245	462	302	50	126
166	133	134	366	386

Solution   76

452	497	527	456	268
521	537	530	166	446
244	115	440	602	799
521	864	513	193	109
462	187	190	783	578

Solution   77

352	367	288	276	87
334	299	276	233	228
157	94	274	357	488
297	546	369	18	140
230	64	163	486	427

Solution   78

395	332	392	298	63
311	296	344	152	377
101	197	296	309	577
294	590	334	146	116
379	65	114	575	347

Solution   79

362	322	376	315	145
348	380	329	131	332
98	141	304	413	564
369	597	333	108	113
343	80	178	553	366

Solution   80

429	442	348	413	103
405	350	389	301	290
261	149	347	342	636
284	628	508	109	206
356	166	143	570	500

Solution   81

417	474	394	482	173
458	474	405	279	324
267	129	388	450	706
380	682	568	117	193
418	181	185	612	544

Solution   82

356	387	391	383	183
388	419	396	154	343
194	110	340	424	632
376	637	430	150	107
386	147	143	589	435

Solution   83

562	490	555	575	213
531	541	578	242	503
258	255	479	497	906
445	862	586	271	231
599	247	197	810	542

Solution   84

341	328	283	326	77
305	292	293	216	249
168	137	271	271	508
237	487	383	89	159
304	111	125	453	362

Solution   85

473	522	532	544	179
513	490	526	252	469
305	199	450	459	837
396	806	616	252	180
563	233	126	743	585

Solution   86

268	243	275	292	132
277	319	262	112	240
106	109	242	304	449
269	434	293	103	111
290	105	138	399	278

Solution   87

265	227	221	220	47
197	224	214	125	220
74	118	196	201	391
200	368	251	67	94
244	43	98	367	228

Solution 88

441	502	457	490	155
461	500	421	258	405
238	161	409	471	766
419	737	583	141	165
486	145	175	685	554

Solution 89

364	379	396	401	115
341	404	363	161	386
160	168	331	344	652
334	592	449	167	113
456	112	116	582	389

Solution 90

477	453	471	444	135
419	419	485	209	448
217	204	396	394	769
371	744	501	212	152
496	160	127	721	476

Solution 91

339	297	376	324	144
343	381	308	119	329
88	151	296	400	545
353	564	325	120	118
357	87	175	517	344

Solution 92

253	261	243	258	70
224	278	219	128	236
94	103	217	244	427
237	393	302	69	84
277	50	104	386	268

Solution 93

365	434	439	377	160
396	431	398	152	398
185	116	355	450	669
409	688	456	149	73
420	106	127	647	475

Solution   94

371	315	388	391	100
330	382	327	155	371
111	211	313	321	609
304	548	393	167	153
449	109	144	531	332

Solution   95

391	469	379	463	193
419	471	440	226	339
275	96	379	427	718
387	674	553	145	136
423	185	144	634	509

Solution   96

281	317	286	271	160
329	350	270	146	220
135	49	263	408	460
342	518	325	38	92
228	81	171	452	383

Solution   97

313	390	351	367	144
374	383	326	192	290
205	92	313	390	565
329	572	444	105	115
344	128	131	511	451

Solution   98

510	470	530	431	214
517	500	490	204	444
184	178	431	586	776
500	862	470	158	165
444	145	234	776	556

Solution   99

401	329	396	412	97
399	334	345	233	324
177	228	327	321	582
249	569	414	172	231
409	175	153	497	401